IOWA BIRD LIFE

PUBLISHED QUARTERLY BY THE

IOWA ORNITHOLOGISTS' UNION

VOL. XVI   MARCH, 1946   NO. 1

OFFICERS OF THE IOWA ORNITHOLOGISTS' UNION

President—Dr. J. Harold Ennis, Mt. Vernon, Iowa
Vice-President—Charles C. Ayres, Jr., Ottumwa, Iowa
Sec'y-Treas.—Miss Lillian Serbousek, 1226 Second St. S.W.,
    Cedar Rapids, Iowa
Librarian—Dr. Warren N. Keck, Cedar Rapids, Iowa
Editor—Fred J. Pierce, Winthrop, Iowa

Executive Council:
    Dr. Robert F. Vane, Cedar Rapids, Iowa
    Mrs. Janet DuMont, Des Moines, Iowa
    Miss Margaret Dorweiler, Cedar Falls, Iowa

The Iowa Ornithologists' Union was organized at Ames, Iowa, February 28, 1923, for the study and protection of native birds and to promote fraternal relations among Iowa bird students.

The central design of the Union's official seal is the Eastern Goldfinch, designated State Bird of Iowa in 1933

Publications of the Union: Mimeographed letters, 1923-1928; 'The Bulletin,' 1929-1930; 'Iowa Bird Life,' beginning 1931.

SUBSCRIPTION RATES: $1.00 a year (includes membership dues). Single copies 25c each. Claims for missing or defective copies of the magazine should be made within six months of date of issue. Keep the Editor informed of your correct address.

EDITORIAL AND PUBLICATION OFFICE
WINTHROP, IOWA

Entered as second-class matter February 9, 1932, at the post office at Winthrop, Iowa, under the Act of March 3, 1879.

THE EASTERN GOSHAWK

"Listed for Iowa as a rare and irregular visitor, in all probability a few of this species are present each year during the fall and winter months."

(From a painting by George Miksch Sutton, reprinted from the 'Wilson Bulletin', December, 1925, facing p. 193.)

bird students to observe as closely as possible any of the long-tailed, rounded-winged hawks seen during the winter months.

SPECIMENS COLLECTED	DATE	LOCALITY	WEIGHT	STOMACH CONTENTS
Male Imm.	10/28/45	Boone Co., Ia.	2½ lb.	Meadow mouse
Male Imm.	10/28/45	Boone Co., Ia.	2 lb., 7 oz.	2 Meadow mice
Female	11/10/45	Forest City, Ia.	3 lb., 6 oz.	Cottontail rabbit

SIGHT RECORDS

Boone County, Iowa, October and December, 1945.

Polk County, Iowa, December 12, 1945.

Ames, Iowa, December 30, 1945, by Wilfred Goodman.

Several conservation officers of the state reported Goshawks in their territories up until January of this year. It is believed that considerable numbers of these birds were in Iowa during the fall and winter of 1945-1946.

GASOLINE ORNITHOLOGY*

By MYRLE L. JONES

Ledges State Park
BOONE, IOWA

There are many ways of enjoying a hobby. Some like to do it the hard way. That seems to be conducive to a more complete relaxation from the routine worries of the usual business.

The writer will be cautious in the beginning and defend himself against any possible criticism of not being able to "take it" by relating that the majority of his winter census trips have been taken largely on foot—many of them involving 8-hour hikes over snow-covered fields in zero weather with the noon meal eaten by one of those typical outdoor fires where the victim alternately freezes and roasts different portions of the anatomy.

Strange, isn't it, how suffering for one's hobby comes back as such pleasant memories? The most nearly fatal experience was that involving the maintenance of a bird feeding-station far from human habitation. In stormy weather this meant a long hike through deep snow. The brush, trees and roughness of the terrain made skis or snowshoes both impractical. Deep snow finally resulted in painful fatigue and leg cramps. The clothing was so wet from perspiration that when finally the open highway was reached, one could not take lightly the danger from freezing on the long ride home in zero weather in the old, canvas-topped, unheated car. One such experience was sufficient and led to easier, if not saner, paths of recreation. It should be mentioned in passing that this ordeal was paid for in full by the record of a Downy Woodpecker banded at the hard-to-reach feeding-station and later captured in Raleigh, North Carolina.

As the bird records pile up year after year, it becomes increasingly obvious that Iowa's bird life has taken to the highways. They have reacted in an interesting way to man's more universal use of gasoline.

The winter habitat of our birds has been improved by several factors. Graveled roads in many parts of the state invite the birds to concentrate along them. These same graveled roads, as well as many side roads and all paved

---

* The illustrations accompanying this article are from 'Iowa Conservationist' and are used through courtesy of the Iowa Conservation Commission.

roads which are kept bare of snow all winter for our convenience in travel, greatly affect our bird life in providing feeding grounds when their usual fields are deeply covered with snow. Not only are the roads and roadsides swept bare, but they are made further attractive to the birds by the more or less continuous trucking of grain, which provides in many places a surprisingly large quantity of available bird feed. All these advantages greatly affect the winter bird population and supplement the list of birds which commonly winter about farm feed-lots.

The most recent Christmas census taken in the vicinity of the Ledges State Park is an example of the opportunities of gasoline ornithology. Of the 25 species of birds observed on this trip, all were seen from the car. While several individual birds were added by side trips on foot into especially inviting bird territory, not one species was thus added to the day's list. This, of course, is not always true and many car stops must be made to be sure of identification, especially of the smaller birds.

If you at first doubt the advantages of this type of bird observation, you might try to recall how many pheasants you have practically run down with your car; then compare the number with those you flushed at close range while on foot.

One of the best known advantages of bird study from the car is the birds' usual fearlessness. Most birds will allow us to approach much closer by car than on foot. Large groups of observers may often study more birds at closer range when in trailer or bus than when on foot. It was the writer's good fortune to work with a group of Boy Scouts who were quite expert in bird observation. On a 600-mile excursion many bird identification tests were passed as their open trailer sped along the highway.

Some cynical person may say that such observations in a moving car are likely to be inaccurate. They could be. Many birds must be put down as unknown unless the trip can be interrupted by frequent but sometimes dangerous stops for closer observation.

"TRY TO RECALL HOW MANY PHEASANTS YOU HAVE PRACTICALLY RUN DOWN WITH YOUR CAR ..."

SWALLOWS ON TELEPHONE WIRES — A COMMON
ROADSIDE SCENE IN LATE SUMMER

Yet there is no particular difference between the probability of errors of the gasoline ornithologist than the observer on foot. In one case the observer moves, in the other the observed moves. What is more disturbing than to barely get the bird glass adjusted on some warbler in the tree top, then have it fly away never to be seen again?

It will be freely admitted that the sparrows and warblers will not usually be studied to advantage from the moving auto. If there is any convenient way to observe the numerous and elusive sparrows and warblers, the method should be made known to the public. One of the best means of learning to identify these birds is through trapping and banding operations, but it can hardly be cited as a convenient method.

Those of you who have studied water birds know of the advantages and limitations of gasoline ornithology. If you are not equipped with webbed feet, you may find it more convenient to spy upon the ducks, geese and some of the other swimmers with telescope while sitting comfortably in your automobile.

Gasoline ornithology in no way takes the place of leisurely trips on foot, especially because the bird song as an identifying factor is often lost to the speeding observer. In this connection it has always been an interesting fact that many birds seem stimulated into vocal activity by the approach of a car. Surely no one has failed to notice how often the Dickcissel on the telephone wire or the Meadowlark on the fencepost burst into song at the passing of the motorist.

These occasional bits of song as well as distinctive color pattern or flight habits can, with practice, be used by all of us to make our business trip or our pleasure drive a more interesting and worthwhile experience.

Observations from the speeding auto or hurried trips to some favorite birding territory may frequently be the only means many of us have at certain seasons of pursuing our avocation. We certainly cannot overlook the importance of the automobile as a means of getting us quickly from one ideal birding area to another.

In the compilation of a large bird list, such as the spring bird census, the "sampling" of the best areas is the only possible solution for a small group seeking to pile up large species lists. In such cases the car takes much of the hard work out of birding or removes the impossible obstacle of covering many miles on foot. Yet we are encouraged to explore likely-looking areas according to our ability, inclination, or the time available.

This time element is one of the strongest points for streamlined bird study in this age of speed. Many quick trips may be made to some favorite bird habitats that we could seldom, if ever, visit on foot.

Some statistics have been worked out on how many birds may be seen per mile, the relative number at different speeds and on paved as compared to gravel or dirt roads, but you will find it more interesting to collect your own data and make your own conclusions. You may find that the observer will locate more birds from a speeding car as a driver than as a passenger, or that there is an optimum speed of travel.

Our spring bird census is most successful if taken when the warbler migration is at its peak. Our warblers, however, have the distressing habit of migrating at that particular period when showers can suddenly appear out of nowhere and sometimes spoil the whole day of birding unless most of it can be done by car. Even this census sometimes proves more successful if taken by car on days when thunder showers are frequent, than a day or so later when the peak of migration has passed even though the weather is ideal.

It is common knowledge to all who operate bird feeding-stations that winter birds feed heavily before a storm. Usually, however, the roadside birds do not respond in the same way. They probably have inland areas from which to feed, and not until the storm has passed and the snow been re-

A TYPICAL WINTER SCENE IN BOONE COUNTY, IOWA
In the recent Christmas bird census taken by M. L. Jones in the vicinity of Ledges State Park, 25 species of birds were seen, all from the automobile.

moved from the roads, do they show up in considerable numbers along the right-of-ways.

One thing that keeps the gasoline ornithologist interested is the occasional observation of the rare or uncommon. The writer has made a number of highway observations over a period of years. A few are listed below, most of them having been published in 'Iowa Bird Life' from time to time either as a note or with the census lists: Red-shafted Flicker, near Guthrie Center, January 30, 1946; Mockingbird, Calhoun County, 1929; Burrowing Owl, Fremont County, 1943; Western Willet, Fremont County, 1943; Krider's Hawk, Fremont County, 1943.

One of the most amusing observations was that of an American Bittern cautiously stalking up the middle of a paved highway that crossed the Missouri River bottoms during the spring flood. He looked as much out of place as a red sweater on a bird hiker. Some of our most interesting bird trips while at Waubonsie State Park were short drives into the picnic area to listen to the Summer Tanagers and Kentucky Warblers. The noise of an approaching car would usually bring forth a burst of song from the Kentucky Warbler. The Summer Tanager usually was not so obliging. A most unusual concentration of Orchard Orioles was observed on a drive from Omaha to Waubonsie State Park. Within less than two hours time 22 Orchard Orioles and 30 Baltimore Orioles were seen.

If it is permissable to digress a trifle from the idea of strictly gasoline-powered transportation, it might interest some to know that even the skeptical Fred J. Pierce once made an almost unheard-of observation of a Great Blue Heron in winter while riding "piggie"-back across the Maquoketa River, the writer furnishing the transportation.

If you are one of those individuals not gyroscopically equipped by nature so that you can drive east while looking north, it will be to the best interests of humanity if you promptly forget this article and concentrate on your driving.

NECROLOGY

Mrs. Frances Davison Ficke, widow of C. A. Ficke, died at her home in Davenport, September 22, 1945. She was born August 31, 1860, the daughter of Abner and Mary Davison, who came to Davenport from New York state when the present city was a pioneer river town. Her father was prominent in the legal profession, and the husband she chose was a young attorney who later became one of the city's leading citizens and a dominent figure in its business and social life.

Mr. and Mrs. Ficke made several trips around the world and visited nearly every foreign country. They were both interested in art and during these trips acquired a large and valuable collection of paintings and other objects of art. This collection was given to the city in 1925 and became the foundation of the Davenport Municipal Art Gallery. Mrs. Ficke was intensely interested in human as well as cultural values. She was one of the founders of the Ladies' Industrial Relief Society and served as its president for 15 years. She gave unsparingly of her energies to the Davenport public library, serving as vice-president of its board for more than 30 years, and always with the desire to broaden its functions and extend its cultural influence. She was a lover of birds and nature and became a member of the Iowa Ornithologists' Union in 1929.

IOWA ORNITHOLOGISTS OF OTHER DAYS
IRA NOEL GABRIELSON

By MRS. H. J. TAYLOR
BERKELEY, CALIFORNIA

The name of Ira Noel Gabrielson is known beyond the limits of this country. His interest in wildlife is neither local nor temporary. It was the great interest of his childhood, and that interest has increased with the years. It embraces all phases of wildlife—its uses and abuses, its value to mankind, its need of conservation and of protection and, wherever possible, its restoration. He sees the whole subject tied up with land, water, and forest, as well as human use and human destruction. The world has needed Ira Gabrielson's comprehensive view of this great subject valuable to all mankind.

Ira Noel Gabrielson was born at Sioux Rapids, Iowa, September 27, 1889. His father was Swedish; he was a farmer in the Drift area of southern Iowa —poorly drained land leaving many small lakes and kettle holes. It was not very good farm land, but it was a delightful place for migrating birds. They came and went with the seasons, probably in great numbers and species. In time the father left the farm and operated a hardware store. He also had interests in a bank in which he took an active part. His mother, of Dutch-English extraction, was a faithful housewife interested in her garden of flowers and vegetables. A garden was the outlet of expression for many a woman of those years.

Ira Gabrielson knew every bird that sang or chirped in his mother's garden. His desire and determination to devote his life to the study of birds and animals caused the father much uneasiness and concern .How anyone could make a living by mere knowledge of birds and animals was beyond the father's comprehension. To him it was a waste of time and a senseless thing. In vain he tried to guide his son into some work whereby he could earn a living for himself; but his efforts were useless. Wildlife, more and more, became the absorbing interest of the son whose mind was definitely made up as to the field of his work. I doubt if he ever thought of "making a living." When the father realized the earnestness and the intensity of his son's attitude toward wildlife he became reconciled to his chosen field of work. Had the father lived long enough, he would have marveled at the attainment of his son.

Ira Gabrielson took his collegiate work at Morningside College, Sioux City, Iowa. He enrolled in 1908. He was delighted to find that he could take training along the line of his special interest. He was most fortunate in having instruction and

DR. GABRIELSON

guidance from a rare teacher. Dr. Stephens is more than a splendid teacher of Biology. His is an instruction that enriches life. Gabrielson specialized in biology and took every available course along that line. He is not, and never was, a library biologist. His greatest pleasure has always been in watching birds and animals and learning from them. It is said that he lets a bear find the sugar in his pocket. We don't mind the sugar but we don't want the bear to get Gabrielson.

While at Morningside College he learned of the great slaughter of animals to get furs for the market. He decided then and there to save the wildlife that had survived. He received his B.A. degree from Morningside College in 1912. In 1936, from Oregon State College, he received the D.Sc. degree and, in 1941, Morningside College conferred on Ira Noel Gabrielson the honorary degree of LL.D. Morningside had recognized in Ira Gabrielson a rare and unusual student, one who would make his mark in the world, yet no one could have predicted all that he has attained. His interest in wildlife and his efforts to preserve it are known throughout the Western Hemisphere. His name has become synonymous with "Wildlife."

On graduating from Morningside College he returned to his home and shortly thereafter married Clara Speer, his sweetheart of high school days. She, too, has become a wildlife enthusiast. They have four daughters.

Dr. Gabrielson has been with the Bureau of Biological Survey since 1915 in various capacities. Since 1935 he has been Chief of the Survey. In 1940 he was made Director of Fish and Wildlife Service (formed by the consolidation of Biological Survey and Bureau of Fisheries).

Many there are who have intense local interest of some phase of wildlife. To protect and increase wildlife requires a much wider view. Dr. Gabrielson has world vision for the needs and welfare of wildlife that belongs to all mankind.

When bird-banding came into being it was widely practiced. There was both pleasure and value in it. In recent years it has been very useful in the wildlife service. In 1938 a cooperator in this field banded 27,076 birds, most of which were migratory waterfowl. There were also, in 1938, 346,056 new birds banded. These represented 431 species that had never before been banded by a wildlife cooperator. Bird-banding brings definite knowledge of birds in certain regions. Chief Gabrielson says that it also emphasized the fact that waterfowl have a very restricted area during the winter months. Through his efforts for the welfare of waterfowl this condition has been improved by preserving the southern marshes—or restoring them—for winter use of waterfowl.

Our government is concerned about its wildlife. The first large refuge was established in 1924. From that time on the movement has made definite progress. In 1929 $10,000,000 was set aside for purchasing and developing waterfowl refuges. In 1940 there were about 14 million acres of land devoted to nearly 300 bird refuges. Birds seem to learn that they are not disturbed on these areas.

Merely being interested in wildlife is not enough to preserve it. Gabrielson begged the country to quit its stream polution and warned, "We are going to find this great land stripped of its wildlife."

A few years ago the Heath Hen vanished. The Trumpeter Swan, the Ivory-billed Woodpecker, the Whooping Crane, and perhaps others, are not entirely out of danger. Much is being done to increase depleted wildlife.

Personally, I was greatly interested when Congress, in 1908, established the Bison Range, the first national field for the vanishing buffalo. This insured the life and the increase of these animals. Everyone knows of the ruthless slaughter of the buffalo merely for its hide. the carcass being left to rot on the plains; later, as the bones were plowed up, they were used to make fences. In 1874, a man who was a good shot, took his vacation with his gun and killed 6,183 buffalo in 60 days. Once we had a million buffalo; now there are but a few thousand.

In 1927 Dr. Gabrielson published an article on the Original Timber Consumer. He reveals the destructive work of the porcupine and says it is second only to the forest fire. In his radio broadcast of January 10, 1936, he said: "If America is not to become the home of the English Sparrow, Norway rats, coyotes, and a few other species that are undesirable in great numbers, there must be something done toward restoring conditions that make life possible for species we want to retain." He used the expression, "Ducks can't nest on a picket fence."

Through his life-long study and thorough knowledge of the subject Dr. Gabrielson is qualified to pass judgment on questions that arise regarding wildlife. The Idaho trapper, after spending some time with him in the field said, "That fellow Gabe has got more information about birds than anyone I ever did meet up with."

Gabrielson is open and above board in all his dealings. His office force are his helpers; he works with them. They are devoted to their leader. Dr. Gabrielson has traveled widely through Canada, Alaska, the Pribilof Islands in the Bering Sea, as well as covering the United States. His writings are many. He has published more than 200 titles, many of which are magazine articles. He has also published a number of books, the most important of which are 'Western American Alpines' (1932), 'Birds of Oregon' (with S. G. Jewett), 1940, 'Wildlife Conservation,' (1941), and 'Wildlife Refuges' (1943).

With the Idaho trapper we say, "That fellow Gabe has got more information about birds (and wildlife) than anyone I ever did meet up with."

THE SPRING CONVENTION AND THE ANNUAL BIRD CENSUS ANNOUNCEMENT

By J. HAROLD ENNIS

We are happy to announce that the 1946 convention of the Iowa Ornithologists' Union will be held in Mount Vernon, Iowa, May 4 and 5. The place of meeting was determined by the members of the Executive Council, as ordered by the business meeting last year in Ottumwa. One interesting historical note is that exactly a half century ago, in 1896, a similar meeting of bird students was held in Mount Vernon on the Cornell College campus, an annual convention of the old Iowa Ornithological Association.

This is the first convention year since the close of the war. With formal transportation restrictions gone, we may expect a large attendance next May. An interesting program is in the making, and you are urged to reserve May 4 and 5 for your trip to Mount Vernon. It is still important to secure an accurate estimate of possible attendance. Secretaries of the various bird clubs are urged to poll their memberships and inform me at Mount Vernon at your earliest opportunity of the number attending.

A second general announcement of importance to the Iowa Ornithologists' Union is the 1946 Spring Bird Count. This is our fourth annual spring census. We are fortunate again that Mr. Myrle L. Jones, Ledges State Park, Boone,

Iowa, has consented to compile your lists. The rules are very simple and may be summarized as follows:

1. Dates—May 10, 11 and 12 (Choose any one of these days).
2. Time—Four-hour minimum.
3. General data desired—Describe territory covered, weather conditions, number of observers, number of hours in the field, number of miles covered on foot and by other means.
4. Census data desired—Names of all species observed, arranged in official A. O. U. order. Number of individuals of each species seen. The letters "C" for counted, or "E" for estimated, may be added to the number of each species.
5. Care should be taken to record only positive identifications.
6. Mail your reports to Mr. Jones, Ledges State Park, Boone, Iowa, before June 1.

Members of the Union are urged to take part in the spring count. This is one of the interesting and valuable projects of our organization that should be supported.

THE 1945 CHRISTMAS BIRD CENSUS IN IOWA

Compiled by FRED J. PIERCE

In the middle west the month of December, 1945, was cold with severe winter conditions prevailing through the entire month. There was considerable snowfall accompanied by sub-zero waves and generally low temperatures. Christmas week was characterized by rigorous weather and adverse conditions unfavorable for field trips and bird observation. However, Iowa bird students have a reputation for hardiness and for liking the weather, regardless of the kind of weather prevailing. This fact is borne out by the large number of observers who took part in the 1945 Christmas census activities.

It is a well known fact that in severe weather winter birds are concentrated into groups and are more easily found and studied than in mild, open weather when they scatter out over a large territory. The Iowa observers made a very thorough survey and a representative listing of the winter bird life, as will be revealed by a careful study of the census tabulation.

Data on place, time, weather and the observers who reported are given below.

1. AMES (Iowa State College campus, along railway and adjacent woods to Agronomy Farm, woods and fields to Horticultural Farm, Brookside Park, and northwest woods to Pammel Woods and golf course; mature deciduous woods 50%, open farmland and fields 25%, campus, etc., 25%): Dec. 30; 8 a.m. to 5 p.m. Cloudy; ground covered with 4-7 in. snow; wind NW, 15-20 m.p.h., temp. 24°; 14 miles on foot. Wilfred Goodman.

The Goshawk was observed in flight and at rest at close range; long tail and short, rounded wings noted. In the Oregon Junco the black head, brown back, pinkish-brown sides, and greater amount of white in tail were carefully studied; its notes were heard.

2. ATLANTIC (Atlantic cemetery, woods along Nishnabotna River, Buck Creek and Indian Creek, Lamb's Lake, Cold Springs State Park, south of city; deciduous woodland 80%, pastures 10%, pine woodland, 10%): Dec. 30; 9:30 a.m. to 5 p.m. Cloudy; streams frozen over; wind NW, 15 m.p.h.; temp. 20 to 30°; total party miles, 62 (40 on foot, 22 in car). Observers in 4 parties. Mr. and Mrs. F. G. Mallette, Dr. and Mrs. Manney Mallette, Charles and Joan

Ruhr, Kenneth Norton, Miss Grace Barnard, Mr. and Mrs. Frank Berry, Bob Mallette, Glen O. Jones, Charles Owen (Atlantic Bird Club).

3. BACKBONE STATE PARK (Delaware County): Dec. 24; 1 to 5 p.m. Overcast and dark; 6 in. snow on ground; wind NE, strong; light to fairly heavy rain all afternoon, freezing as it fell; temp. 22° at start, 25° at return; about 2 miles on foot, 40 by car; trip included an auto trip from Winthrop to the park and return, with roadside birds included in the census. Observers together. Mr. and Mrs. M. L. Jones, F. J. Pierce.

4. CEDAR FALLS (Cedar River, Beaver and Snag Creeks, Josh Higgins Park: river-bottoms 50%, upland forest 20%, savannas 20%, farmland 10%): Dec. 28; 8 a.m. to 12:15, 1:45 to 4:45 p.m. Overcast; ground snow-covered: open patches in river; wind W, 0-5 m.p.h.; temp. 20° -26°; total miles, 20 by car, 12 on foot. Observers together. Martin L. Grant, Maybelle Brown, Mrs. Oren Paine, Mrs. Arthur Lynn, Gordon Grant (Cedar Falls Audubon Society).

5. CEDAR RAPIDS (Ellis Park, Cedar Lake, Mound Farm woods, Chain Lakes, river road): Dec. 30; 9 a.m. to 5 p.m. Cloudy, visibility poor; deep snow, and trees covered with ice; wind NE; temp. 22° to 26°; 10 miles by car, 5 miles on foot. Dorothy B. Hayek, Lillian Serbousek, Dr. and Mrs. Robert Vane, Myra Willis.

6. CLARION (Clarion Evergreen Cemetery, down White Fox drainage ditch south of cemetery for ½ mile on foot, auto ride 5 miles east of town, north and back to town; feeding-station in town): Dec. 28; 9 a.m. to 1 p.m. Cloudy; 3-6 in. snow on ground; wind S; temp. 19° at start, 22° at return. Observers together. Mrs. W. C. DeLong, Richard DeLong.

7. DAVENPORT (Giddings woods, Credit Island, McMannus woods, Blackhawk Creek, Holy Family Cemetery, Stubbs woods, Rowes Creek, fairgrounds, Clader St. woods and creek, along Rock Island Ry. tracks to Harbor Road): Dec. 24; 8 a.m. to 4:30 p.m. Cloudy; heavy snow which fell previous night; began raining at 3 p.m.; temp. 20° at start, 32° at return; about 17 miles on foot. Jim Hodges.

The Snow Geese were observed at 100 ft. with a 5-power glass. When first seen they were resting on the ice on the river beside an open hole. Plumage was snow white, and when in flight black wing-tips were noted.

8. DAVENPORT (Duck Creek Park, along creek to Devil's Glen, Credit Island and nearby shore area along Mississippi River, through fields to Fairmount Cemetery): Dec. 25; 7:40 to 10:50 a.m., 11:45 a.m. to 4 p.m. Cloudy; 4-6 in. snow on ground; creeks, ponds and harbor frozen; river frozen except for small area at head of island; snow falling until 1:30 p.m.; wind W; temp. 33° at start, 29° at return; about 12½ miles on foot. C. F. Mueller.

9. DES MOINES (Dove and Kinglet woods along Beaver Creek, both sides of Des Moines River from Crocker Woods to Lovington, Sycamore Park, Fisher's Lake, Morningstar, Pine Hill Cemetery, Ashworth Park, Terrace Road and vicinity, Audubon Society Sanctuary: open woodlands, along streams, roadsides, cornfields near woodlands): Dec. 23; 8 a.m. to 4:30 p.m. Cloudy, snowing in p.m.; 3 in. snow on ground: all streams and lakes frozen: wind SE, 15 to 25 m.p.h.; temp. 14° at start, 19° at return; total party miles, 117½ (37½ on foot, 80 by car). Ten observers in 6 parties. Albert C. Berkowitz, Woodward Brown, Mrs. W. G. DuMont, Olivia McCabe, Elizabeth Peck, Mrs. Harold R. Peasley, Bruce F. Stiles, Jack W. Musgrove, Irene M. Smith, Mrs. Toni R. Wendelburg.

The record for Broad-winged Hawks was made by Olivia McCabe, Elizabeth Peck and Mrs. W. G. DuMont. Red-breasted Nuthatch and Yellow-bellied Sapsucker were observed at a feeding-tray during Christmas week.

(Continued on page 16)

Species	1. Ames	2. Atlantic	3. Backbone State Park	4. Cedar Falls	5. Cedar Rapids	6. Clarion	7. Davenport	8. Davenport	9. Des Moines	10. Dubuque	11. Keosauqua	12. Ledges State Park	13. Mt. Pleasant	14. Mt. Pleasant	15. Mt. Vernon	16. Ottumwa	17. Sioux City	18. Sioux City to Hornick	19. Waterloo to North English	20. Woodward
Snow Goose																				
Mallard																				
Blue-winged Teal																				
American Golden-eye																				
American Merganser																				
Goshawk																				
Sharp-shinned Hawk																				
Cooper's Hawk																				
Red-tailed Hawk																				
Red-shouldered Hawk																				
Broad-winged Hawk																				
Rough-legged Hawk																				
Bald Eagle																				
Marsh Hawk																				
Sparrow Hawk																				
Prairie Falcon																				
Bob-white																				
Ring-necked Pheasant																				
Wilson's Snipe																				
Ring-billed Gull																				
Mourning Dove																				
Screech Owl																				
Great Horned Owl																				
Barred Owl																				
Long-eared Owl																				
Short-eared Owl																				
Saw-whet Owl																				
Belted Kingfisher																				
Flicker																				
Pileated Woodpecker																				
Red-bellied Woodpecker																				
Red-headed Woodpecker																				
Hairy Woodpecker																				
Downy Woodpecker																				
Horned Lark																				

Stations:

1. Ames
2. Atlantic
3. Backbone State Park
4. Cedar Falls
5. Cedar Rapids
6. Clarion
7. Davenport
8. Davenport
9. Des Moines
10. Dubuque
11. Keosauqua
12. Ledges State Park
13. Mt. Pleasant
14. Mt. Pleasant
15. Mt. Vernon
16. Ottumwa
17. Sioux City
18. Sioux City to Hornick
19. Waterloo to North English
20. Woodward

Species:

Blue Jay
Crow
Chickadee
Tufted Titmouse
White-breasted Nuthatch
Red-breasted Nuthatch
Brown Creeper
Winter Wren
Robin
Bluebird
Golden-crowned Kinglet
Northern Shrike
Starling
English Sparrow
Western Meadowlark
Red-winged Blackbird
Rusty Blackbird
Bronzed Grackle
Cardinal
Purple Finch
Redpoll
Pine Siskin
Goldfinch
Vesper Sparrow
Slate-colored Junco
Oregon Junco (subsp. ?)
Tree Sparrow
Harris's Sparrow
Song Sparrow
Lapland Longspur
Snow Bunting
Number of Species
Number of Observers

*See data under station in body of article.
Total Iowa List ....66 species.

10. DUBUQUE (Linwood and Mount Calvary Cemeteries, Eagle Point Park, Mississippi River sloughs in Wisconsin, Catfish Creek area from Rowan St. to Mississippi River; pine woodland 15%, deciduous woodland 35%, untilled fields 25%, river sloughs 25%): Dec. 30: 8:30 a.m. to 12:30, 1:30 to 4:30 p.m. Overcast with light snow for 2 hours; 8 in. old, crusted snow on ground; all water frozen except below dam and swift-flowing current in creek; wind NW, 2-4 m.p.h.; temp. 25°-30°; total miles, 19 (13 on foot, 6 by car). Observers in 2 parties. Jess Crossley, George Crossley, Mr. and Mrs. James Dockal, Ethan Hemsley, Henry Herrmann, Ed Heuser, David Reed, Ival Schuster (Dubuque Bird Club).

11. KEOSAUQUA (and immediate vicinity): Dec. 26; 9 a.m. to 12, 1 to 3 p.m. Cloudy, very hazy and difficult to see birds; ground snow-covered; frozen ice crust on trees, etc.; temp. 6° at start, 18° at noon, 12° at return. Warren N. Keck.

12. LEDGES STATE PARK (Boone Co.): Dec. 31; 9 a.m. to 5 p.m. Clear after 10:30 a.m.; 6 in. snow on level, loose snow drifting all day on uplands; high wind, and snow flurries in a.m.; temp. 12° at start, up to 20° at warmest point, 12° at return, a bitterly cold day; 3 miles on foot and lengthy drive by car. M. L. Jones.

The Vesper Sparrow was observed five different times, finally at 25 ft. with and without glass. Its right wing was drooping as though it had been broken, but its flight seemed to be normal. It was near a neglected orchard and was not with any other birds.

13. MOUNT PLEASANT (South from Saunder's Grove to Big Creek): Dec. 26; 9 a.m. to 2 p.m. Clear and still; 8 in. snow on ground; temp. 20° to 30°; about 8 miles on foot. N. L. Cuthbert.

14. MOUNT PLEASANT (City Park and southwest, old stone quarries, wooded hills and small streams one of which had open water, two cornfields): Dec. 28; 1 to 4 p.m. Cloudy and foggy; ground snow-covered; no wind; temp. 31°; 5 miles on foot, 1 mile by car. Roy Ollivier, Joe Schaffner.

The Wilson's Snipe was seen along an open stretch of water in a small creek just below the city septic tank and probably kept unfrozen by this. This species has remained there for some time and as many as four have been seen at once. Two years ago two Wilson's Snipes wintered along the same creek.

15. MOUNT VERNON (north of town 3 miles, including old golf course; south of town 6 miles to Cedar River, and through length of Dark Hollow in Palisades State Park; open farm land 60%, town 5%, deciduous timber land through Park 35%): Dec. 29; 7½ hours. Heavy fog and poor visibility. J. Harold Ennis.

16. OTTUMWA (Community Gardens, Hamilton Park, Memorial Park, Pruitt's Lake and vicinity): Dec. 23; 9 a.m. to 1, 2 to 4 p.m. Cloudy; ground covered with ice and drifted snow; strong wind, 15-25 m.p.h., and heavy snowstorm in p.m.; temp. 18° at start, 16° at return; 15 miles on foot. Billie Hoskins, Mary Wood, Norman Crowe, Jane Wood, Pearle Walker, Marilyn Watterson.

17. SIOUX CITY (War Eagle's Monument Park woods, confluence area of Missouri and Big Sioux Rivers, Lower Riverside woods, Riverview Park, Riverside Park, area along road at foot of loess hills lying between Stone Park and Riverside, Stone Park, Plum Creek, Griffen's woods, Highview Golf Course area, North Side, West Side; hilly woodland, mature deciduous timber,

	1. Ames	2. Atlantic	3. Backbone State Park	4. Cedar Falls	5. Cedar Rapids	6. Clarion	7. Davenport	8. Davenport	9. Des Moines	10. Dubuque	11. Keosauqua	12. Ledges State Park	13. Mt. Pleasant	14. Mt. Pleasant	12. Mt. Vernon	16. Ottumwa	17. Sioux City	18. Sioux City to Hornick	19. Waterloo to North English	20. Woodward
Blue Jay		6	6	6	16	3	16	19	11	12	2	6	6	6	8	30	27	6	6	1
Crow	11	20	12	150	33	6	6	19	165	1287		21	19	18	4	56	11	150	60	8
Chickadee	48	78		30	23	6	62	19	10	67		28	26	2	2	25		3	5	39
Tufted Titmouse		1	1		2				10	7	4	1	1	1	2	18	6		1	
White-breasted Nuthatch	16	32		18	10	1	13	3	18			1	6	1			4			6
Red-breasted Nuthatch																				
Brown Creeper	3			1			2	1		15	4	7					7	4		
Winter Wren																				
Robin																				
Bluebird																				
Golden-crowned Kinglet	21											21	16							
Northern Shrike		1				16	3		1	2	25	1			11	35		11	60	2
Starling	10	27				29	19	1000	1	91	52	1	16	16	18	360	360		30	6
English Sparrow	128	688		60		2	66	268	551	35		76	16	8						
Western Meadowlark	2	16			21				3			1								16
Red-winged Blackbird		3																		
Rusty Blackbird																				
Bronzed Grackle										2	8				8	20	5		3	
Cardinal	8	36	75	29	21	5	9	17	77	9	4	14	10	17		4	6		9	6
Purple Finch									119	22			60	4	8		5	11		72
Redpoll														21						
Pine Siskin												11								
Goldfinch	66	40	4	60	12	8	81	32	213	21	20	1	13	21	8	10	136	11	9	48
Vesper Sparrow	2	1							362	29	20	50	3		5	150	16	2	4	
Slate-colored Junco	14			1	43		1	1	1	16	11	38	3		8	85	16	150	200	
Oregon Junco (subsp. ?)	1	22				3			1	1	1									
Tree Sparrow		12	75	65	50	60	2	131	131				3	19		1				
Harris's Sparrow																				
Song Sparrow	1			2	2															
Lapland Longspur																				
Snow Bunting																				
Number of Species	25	30	15	20	23	18	26	22	35	26	18	25	21	18	15	23	27	15	17	18
Number of Observers	1	10	31	5	6	2	1	1	101	91	1	1	1	2	1	6	13	1	1	1

*See data under station in body of article.
Total Iowa list ....66 species.

10. DUBUQUE (Linwood and Mount Calvary Cemeteries, Eagle Point Park, Mississippi River sloughs in Wisconsin, Catfish Creek area from Rowan St. to Mississippi River; pine woodland 15%, deciduous woodland 35%, untilled fields 25%, river sloughs 25%): Dec. 30: 8:30 a.m. to 12:30, 1:30 to 4:30 p.m. Overcast with light snow for 2 hours: 8 in. old, crusted snow on ground: all water frozen except below dam and swift-flowing current in creek: wind NW, 2-4 m.p.h.; temp. 25°-30°; total miles, 19 (13 on foot, 6 by car). Observers in 2 parties. Jess Crossley, George Crossley, Mr. and Mrs. James Dockal, Ethan Hemsley, Henry Herrmann, Ed Heuser, David Reed, Ival Schuster (Dubuque Bird Club).

11. KEOSAUQUA (and immediate vicinity): Dec. 26: 9 a.m. to 12, 1 to 3 p.m. Cloudy, very hazy and difficult to see birds: ground snow-covered; frozen ice crust on trees, etc.; temp. 6° at start, 18° at noon, 12° at return. Warren N. Keck.

12. LEDGES STATE PARK (Boone Co.): Dec. 31: 9 a.m. to 5 p.m. Clear after 10:30 a.m.; 6 in. snow on level, loose snow drifting all day on uplands: high wind, and snow flurries in a.m.; temp. 12° at start, up to 20° at warmest point, 12° at return, a bitterly cold day: 3 miles on foot and lengthy drive by car. M. L. Jones.

The Vesper Sparrow was observed five different times, finally at 25 ft. with and without glass. Its right wing was drooping as though it had been broken, but its flight seemed to be normal. It was near a neglected orchard and was not with any other birds.

13. MOUNT PLEASANT (South from Saunder's Grove to Big Creek): Dec. 26: 9 a.m. to 2 p.m. Clear and still: 8 in. snow on ground; temp. 20° to 30°; about 8 miles on foot. N. L. Cuthbert.

14. MOUNT PLEASANT (City Park and southwest, old stone quarries, wooded hills and small streams one of which had open water, two cornfields): Dec. 28: 1 to 4 p.m. Cloudy and foggy; ground snow-covered; no wind; temp. 31°; 5 miles on foot, 1 mile by car. Roy Ollivier, Joe Schaffner.

The Wilson's Snipe was seen along an open stretch of water in a small creek just below the city septic tank and probably kept unfrozen by this. This species has remained there for some time and as many as four have been seen at once. Two years ago two Wilson's Snipes wintered along the same creek.

15. MOUNT VERNON (north of town 3 miles, including old golf course: south of town 6 miles to Cedar River, and through length of Dark Hollow in Palisades State Park: open farm land 60%, town 5%, deciduous timber land through Park 35%): Dec. 29; 7½ hours. Heavy fog and poor visibility. J. Harold Ennis.

16. OTTUMWA (Community Gardens, Hamilton Park, Memorial Park. Pruitt's Lake and vicinity): Dec. 23: 9 a.m. to 1, 2 to 4 p.m. Cloudy; ground covered with ice and drifted snow; strong wind, 15-25 m.p.h., and heavy snowstorm in p.m.; temp. 18° at start, 16° at return; 15 miles on foot. Billie Hoskins, Mary Wood, Norman Crowe, Jane Wood, Pearle Walker, Marilyn Watterson.

17. SIOUX CITY (War Eagle's Monument Park woods, confluence area of Missouri and Big Sioux Rivers, Lower Riverside woods, Riverview Park, Riverside Park, area along road at foot of loess hills lying between Stone Park and Riverside, Stone Park, Plum Creek, Griffen's woods, Highview Golf Course area, North Side, West Side; hilly woodland, mature deciduous timber,

pine and cedar growth 50%, river country and wooded lowland 25%, open hills 15%, town 10%): Dec. 23; 9:30 a.m. to 5 p.m. Totally overcast; 3½ in. snow on ground; all river open but with ice crusts near shores; all creeks frozen; wind SE, 25 to 30 m.p.h., with light snow from 11:10 a.m. to 12:45 p.m.; temp. 6° to 20°; 8 miles on foot, 36 by car. Observers in 8 parties. Don Bushar, Mrs. Marie Dales, Dr. J. E. Dvorak, Mrs. E. A. Emery, Ethel Hackett, Jayne Hemmingson, Jean L. Laffoon, Mr. and Mrs. H. T. Lambert, Zell C. Lee, Mrs. J. L. Schott, Bertha Wellhausen, Carl Wellhausen (Members and guest, Sioux City Bird Club).

18. SIOUX CITY TO HORNICK (an auto trip of 53 miles in Woodbury County, via Missouri River bottoms and return): Dec. 31. Clear and bright; light wind blowing fine snow of previous night's fall across roads and fields; temp. 10°-11°. Wm. G. Youngworth.

19. WATERLOO TO NORTH ENGLISH (100-mile auto trip): Dec. 25; 9:30 a.m. to 12:30, 3:30 to 4:30 p.m. Cloudy, with misty rain; 4 in. snow on ground; one-eighth in. ice on ground at Waterloo, three-eighths at North English; temp. about 24°; also 2 miles on foot. M. L. Jones.

20. WOODWARD (area 4 to 5 miles SE of town; timber, valleys and weed patches): Dec. 25. Heavy clouds, and snow in air at times; 8 in. snow on ground; wind NW, 12 m.p.h.; 4 miles on foot, 12 by car. Richard A. Guthrie.

GENERAL NOTES

Evening Grosbeaks at Ames.—On November 4, 1945, I had distinguished visitors. I saw them first at the bird-bath, four strangers—the distinctive male and three more somber-colored companions. I knew at once they were grosbeaks, but it was not until a number of other bird students had responded to my urgent call over the phone that I realized fully what an honor had been vouchsafed us. The grosbeaks were viewed by Dr. and Mrs. George Hendrickson, Mrs. Carr, Miss Margaret Murley, Mrs. Mary Quam, and Mrs. W. P. Battell.

Breeding in Manitoba and the extreme northern parts of the United States east of the Rockies, the Evening Grosbeak's winter migrations depend more on the food it can find than on the weather, and so, like the Bohemian Waxwing, it does not have a fixed migration route. Ordinarily it travels in rather small groups (a half dozen or so up to 60), but on several occasions very large numbers have appeared in northern United States, and well beyond the usual haunts of the species. Its food is said to consist of seeds from box-elder, sugar maple, ash, and from the fruits of the hackberry, highbush cranberry, mountain-ash, juniper, cedar and sumac. While at my home they all flew to the feeding-shelf and ate the sunflower seeds they found there. They seemed tame and unafraid.

Various reports on its vocal attainments are not flattering and all mention one peculiarity—its song, in ascending notes, ceases abruptly as though the bird had lost its breath. Peterson says "the note sounds like a ringing, glorified chirp of a House Sparrow." During the two hours or more the birds were here they were observed very carefully by the group of bird students mentioned above, who decided that we had a male, a female and two immature grosbeaks. Their plumage was in prime condition. They paid no attention to the group assembled in the room a short distance from the window.—MRS. F. L. BATTELL, Ames, Iowa.

RECENT BIRD BOOKS

THE RING-NECKED PHEASANT AND ITS MANAGEMENT IN NORTH AMERICA. Edited by W. L. McAtee. (The American Wildlife Institute, Washington. D.C., 1945; cloth, pp. i-xi + 1-320, 2 colored plates, 31 halftones, 12 figures. 19 tables; price $3.50).

We draw heavily on the Foreword by Mr. McAtee, the eminently able and careful Editor, in this review. The foundation of this book is a report on the ecology, life equation and management of the Ring-necked Pheasant in Ohio. Supplementary are chapters relating to seven other important centers of pheasant distribution across the continent and one on artificial propagation of the bird. Most of these have been prepared by employees of the federal Fish and Wildlife Service and are largely products of Cooperative Wildlife Research Units, jointly sponsored by that Service, certain State agricultural colleges, including Iowa State College, the respective State Conservation departments and the American Wildlife Institute. "The Historical Introduction" is by ex-Senator Frederic C. Walcott of Connecticut, an outstanding leader in the conservation of natural resources. Jean Delacour, President, International Committee for Bird Preservation, furnished the chapter on "Classification and Distribution of the Game, or True, Pheasants."

The management principles for the maintenance of conditions favorable for pheasants are of broad application. Management must conform to the system of agriculture in a region.

Iowa's part in the book is prominent . The book is dedicated to the late Professor Howard Marshall Wight, who studied toward the degree of Doctor of Philosophy in Zoology at Iowa State College in 1926-27 while Professor J. E. Guthrie had charge of wildlife conservation. Professor Wight, and two of his advanced students at Michigan State University, Drs. Paul D. Dalke and P. F. English, prepared the chapter on "The Pheasant in Michigan." Dr. Logan J. Bennett. Ph. D. Iowa State College, 1937, now Leader, Pennsylvania Cooperative Wildlife Research Unit, wrote "The Pheasant in Pennsylvania and New Jersey. Dr. Paul L. Errington, Research Associate Professor in Economic Zoology, Iowa State College, supplied "The Pheasant in Northern Prairie States," including Iowa, in the absence in the Armed Services of Dr. Thomas G. Scott, Leader, Iowa Cooperative Wildlife Research Unit, and Dr. Thomas S. Baskett, who received his Ph. D. at Iowa State College, 1942, with a thesis on the Production of the Ring-necked Pheasant in North-Central Iowa. Dr. H. Elliott McClure, Ph. D., Iowa State College. 1941, is co-author with Dr. Ward M. Sharp of "The Pheasant in the Sandhill Region of Nebraska." Fourteen of 79 literature citations are to articles by members of the Cooperative Wildlife Research Unit at Iowa State College.

Although many problems in pheasant management are yet to be solved, this book is the latest, most extensive and scientifically authoritative volume in the field and a monument to the efforts of the numerous cooperating agencies.—George O. Hendrickson.

* * * * *

CHECK-LIST OF THE BIRDS OF NEBRASKA, by F. W. Haecker, R. Allyn Moser, and Jane B. Swenk, (Nebr. Ornith. Union, Omaha, Nebr., 1945; wrappers, pp. 1-44; price, 50c).

The last annotated check-list of Nebraska birds was published in 1920. For some time there has been need for a new list, since Nebraska has a large group of serious bird students ready to make use of it. The new list includes all published and unpublished records known to the present time, a total of 472 species and subspecies. A thorough study has been made of subspecific forms, and the list as now published is an accurate representation of these perplexing subspecies as they are now understood. The list is reprinted from 'Nebraska Bird Review', semi-annual magazine of our neighboring state bird society. The Check-list should have a good deal of interest for Iowa bird students.—F.J.P.

* * * * *

AN ANNOTATED BIBLIOGRAPHY OF SOUTH DAKOTA ORNI-THOLOGY, by T. C. Stephens (published privately, Sioux City, Iowa, 1945; wrappers, pp. i-iv — 1-28, lithoprinted; price, $1.)

Dr. Stephens has been working on this South Dakota bibliography for more than 25 years. In his preface he states that he was unable to inspect, first hand, many of the earlier papers and suggests there may be omissions. This may be true, but running through the book we are impressed by the large number of titles that he has indexed, which is an indication of the thoroughness of the work. The first entry is for 1858, and the titles continue down through the years to 1944, each title followed by an annotation regarding the species mentioned in the article with dates and other information. Dr. Stephens has performed a great service for bird students of South Dakota and adjacent regions.—F.J. P.

MEMBERSHIP NEWS

Paul A. Stewart, a member of the Union living at Leetonia, Ohio, visited the Editor on December 24. He was on his way home from a business trip to Nebraska, and had taken a bird census at Valentine, Nebraska, on a previous day. Mr. Stewart is a farmer, ornithologist and book collector. We are always glad to meet our out-of-state members.

Mrs. Ella L. Clark, of Burlington, has donated to the Union, a nearly complete file of 'Iowa Bird Life'. We are glad to receive this set as it gives us a number of issues of which our stock is very low. We have only four complete sets of 'Iowa Bird Life' on hand, which means that of certain issues only four copies are in stock. We hope that none of our members will destroy their files of the magazine, and suggest that any unwanted copies be returned to the Union for use in filling requests for back numbers.

Dr. and Mrs. F. L. R. Roberts, of Spirit Lake, enjoyed a few days of vacation in and about Boulder, Colorado, in December. "Doc" said they saw and compared four kinds of juncos, all in one flock, and that they had "some excellent chances at Lewis's Woodpecker, House Finch, etc., with 4-inch movie lens, but found when we got home that the camera was out of order."

Robert A. Pierce, formerly of Nashua, Iowa, has received an appointment to the U.S. Fish & Wildlife Service with headquarters at Atlanta, Georgia. In February he was working on an assignment in Arkansas, a member of a party making topographical maps of the Arkansas River bottoms. He reported seeing many interesting birds in this region, including Mallards, Blue Geese, Bewick's Wren and various others.

Our Vice-President, Charles C. Ayres, Jr., had a fine trip to California from December 12 to January 23. He attended one meeting of the Los Angeles Audubon Society and one of the Southwest Bird Study Club, and participated in one of the regular hikes of the Los Angeles organization. He also had dinner with Bert Harwell and went with him to one of his lec-

tures, operating the projector for him. He reports that the bird lovers of
California entertained him royally and took him around to see as much of
their bird life as was possible in the time available. He estimates that
they drove him from 1000 to 1500 miles on the bird trips. On trips through
the desert he saw Vermilion Flycatcher, Verdin, Mearn's Gilded Flicker,
LeConte's Thrasher and Western Gnatcatcher. Along the shores of the
ocean and in the sloughs near by, he saw hundreds of Willets. Marbled God-
wits. Hudsonian and Long-billed Curlews. Sanderlings, Black-bellied Plov-
ers and others. On the water there were Sooty Shearwaters, Surf and
American Scoters, three kinds of loons, and all three of the cormorants.
On a trip north of Los Angeles he had the privilege of seeing five of the
rare White-tailed Kites, and in the redwoods around Santa Cruz, he saw
the Steller's Coast Jay and Varied Thrush. Another treat was the sight
of an estimated thousand Mountain Bluebirds on a relatively small plowed
field.

Mr. Ayres says: "Probably the top feature of our birding trips was a
journey 11 miles up a mountain trail where, at an elevation of nearly 5000
feet, we watched five California Condors soaring over the mountains. Observation was made both by binoculars and spotoscope. There are some 40
of these birds still known to be alive and, as I understand it. California is
the only place where they are found. The Condor trip entailed some danger
from my point of view. I think you would agree on this if you could see
the mountain road which we traveled. It was so narrow that one could
not have walked beside the car; there was a steep mountain on one side
and a thousand-foot drop on the other."

THE FIVE-YEAR INDEX. Our December issue contained the cumula-
tive five-year index, which was the result of several persons' work. Index-
es to periodicals are a necessary and important feature if the publication is
to have permanent reference value. The amount of work required in their
compilation, however, is sometimes not understood or appreciated. The
Editor began work on the five-year index—the third one he has compiled
—early in October, and worked on it during many half days during that
month (exact hours were not recorded). The data were first set down on
3 by 5 index slips, and later arranged and typed in manuscript form for the
printer. The manuscript was completed and sent to the printer on November 1, but through an unfortunate combination of circumstances and delays,
the printer was unable to get the galley proof to the Editor until January
10.

Mrs. W. C. DeLong of Clarion accepted the task of checking the proof
against the magazine. Using copies of 'Iowa Bird Life' for reference, she
looked up every entry in the index to make sure that it was correct as to
title, bird species, volume number and page, respectively. This was no
small undertaking, and she very generously contributed 21 hours and 25
minutes of her time to this work (January 11-20). Reva Pierce read the
proof back against the manuscript and thus furnished a double check on
such details as spelling and correct transcription. The corrected galley
proof was sent to the printer on January 24, after which a page-proof was
furnished. The page-proof was in turn carefully checked against the galley to make sure that no items had been omitted or shifted in columnar
position. The corrected page-proof was returned to the printer on February 1. Then the index was ready for printing. We mention these steps in
the production of the index so that our readers will have at least a slight
idea of how much work goes into the making of an accurate index.

CPSIA information can be obtained
at www.ICGtesting.com
Printed in the USA
LVHW070932110623
749445LV00018B/214